Observing and Classifying Matter

Matter 2
Solids 4
Measuring Solids 6
Liquids 8
Air and Other Gases 10

Harcourt
SCHOOL PUBLISHERS

Orlando   Austin   New York   San Diego   Toronto   London

Visit *The Learning Site!*
www.harcourtschool.com

Matter

Matter is what all things are made of.
All matter takes up space.
All matter has mass.
These are properties of matter.

Size is also a property of matter.
Color and shape are other properties.

Solids

A solid is a form of matter.
Some solids are hard. Some are soft.
Solids are different sizes and colors.
Solids have different textures.

All solids have a shape of their own.
You can do things to change the shape.
You can bend a solid or break it or cut it.
But the shape does not change on its own.

Measuring Solids

balance

You can measure the mass of a solid.
A balance helps you measure mass.

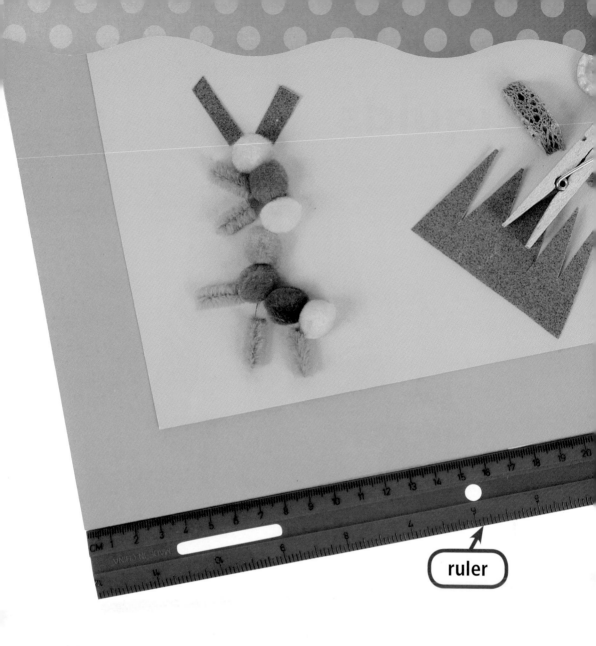

ruler

You can measure the size of a solid.
A ruler helps you measure size.
A ruler may measure in centimeters.
Or, it may measure in inches.

Liquids

A liquid is a form of matter.
Liquids do not have their own shape.
They take the shape of their container.

measuring cup

You can measure the volume of a liquid.
A measuring cup measures volume.
It may measure volume in milliliters.
Or, it may measure volume in ounces.

Air and Other Gases

A gas is a form of matter.
A gas has no shape of its own.
A gas fills its entire container.

Air is made up of gases.
You can not see or smell air.
But you can see what air does.

Vocabulary

centimeter, p. 7

gas, p. 10

liquid, p. 8

mass, p. 2

matter, p. 2

milliliter, p. 9

property, p. 3

solid, p. 4

texture, p. 4

volume, p. 9

Think About the Reading

1. What questions do you have after reading this book?

2. How can you find the answers to these questions?

Hands-On Activity

1. Put five balloons in a paper bag.

2. Blow up the balloons with air. Put them back in the bag. What do you observe? What property of matter does this show?

School-Home Connection

Tell a family member what you have read about changes in matter. Talk with the family member about the ways matter changes when food is prepared.

BPP05

SCHOOL PUBLISHERS
www.harcourtschool.com

ISBN 0-15-343830-4

9 780153 438301